TESTIMONIALS OF IMPACT

The more you know, the more you will grow—but knowledge alone is not enough. Growth requires action. It requires watering the seeds of knowledge, nurturing them with effort, persistence, and faith. This book you hold in your hands is not just a collection of words; it is a blueprint for transformation. Jabez doesn't just share knowledge—he shares the power of application. His journey proves that your life can change when you are bold enough to apply what you know.

I am proud to call Jabez my brother and to have been a witness to his growth from the start. This is not just a introduction—it is a celebration of a man who turned his dreams into reality and is now helping others to do the same. Read this book not just to know but to grow. Plant your seed, water it with action, and watch the extraordinary life you can create.

With gratitude and purpose,
Billionaire PA
Entrepreneur | Motivational Speaker | World Changer

Like a lot of you probably reading these words, I became friends with Jabez through his social media platforms. His positive energy towards life and growing food is contagious! Whether you're just starting out on your food growing journey or a master grower I can guarantee you'll take away lots of great knowledge from this book. His approach to mixing modern day technology and old school methods will guide you towards having a thriving garden no matter where you live in the world.

> *"Abundance is all around us,
> everything we need is already here
> waiting for us to use it wisely!"*
> —*Jabez*

Anthony Coletti
CEO/Co-Founder of Grow Inn Homes

Many people know Jabez for his transformative work in the health and wellness space, but fewer know the side of him I first encountered—as an incredible football player and teammate. Jabez and I met at Florida International University. As a student-athlete, I often sought his advice on balancing academics, athletics, and life's challenges.

As someone who now provides protection for politicians, activists, celebrities, and athletes, I find myself recommending Jabez's philosophy and insights to many of the people I meet. He's someone I trust—not just because of our history but because of the sincerity and effectiveness of his work.

This book is more than a guide—it's an invitation to learn from someone who has walked the walk, someone whose experiences can help you grow into a more fulfilling life. Most importantly, it's an opportunity to connect with a person who genuinely cares about making a difference. To everyone reading this book, I hope it reignites your passion for growing natural food and reminds you that every mistake is just a step closer to success. Let this be the spark that inspires you to embrace the journey, take action, and grow for yourself and your loved ones.

Alexander Bostic
Executive Protection
Watchman Security Services

Copyright © 2025 Jabez el Israel

All rights reserved. No part of this publication may be reproduced, distributed, or transmitted in any form of by any means, including photocopying, recording, or other electronic or mechanical methods, without the prior written permission of the publisher, except in the case of brief quotations, embodied in reviews and certain other non-commercial uses permitted by copyright laws.

Editor: Janine Hernandez, The Book Publishing Academy
Website: www.janinehernandez.com

Amazon Book Coach: Tony Shavers III Consulting
Website: www.tshavers.com

Graphic Designer: Mal Graphics
Instagram: @malgrfx

Interior Formatter: Wild Words Formatting
Website: www.wildwordsformatting.com

Printed in the USA

EBook ISBN: 979-8-9861241-4-8
Paperback ISBN: 979-8-9861241-5-5

DEDICATION

This book is dedicated to my family,
Nicole, Supreme, Shaniya, Budda, Jacinda and Mekhi.
You guys are my why and I love you deeply!

TABLE OF CONTENTS

Foreword by Kenneth Ott — 1

Introduction: The Hidden Cost of Convenience — 7

1. The First Step — 13

2. Understanding Your Environment: Growing Zones — 27

3. Water and Sunlight: Life's Essentials — 37

4. Soil – The Foundation of Growth — 45

5. Seed vs. Seedlings – Choosing Your Start — 63

6. Propagation, Air Layering, and Grafting – Multiplying Plants Without Seeds — 71

7. Fertilizer – Nourishing the Roots — 81

8. Harvesting – When and How to Reap What You Sow — 89

9. Healing Sick Plants – Don't Be So Quick to Give Up — 97

10. The More We Grow, The More We'll See Everything We Need Is Free! — 111

Acknowledgments — 121

FOREWORD

Something is very wrong with how we get our food today.

I grew up in Queens, New York in the 1980s. Back then, almost everything we ate came in cardboard boxes and plastic packages. The only garden I ever saw was in my cousins' backyard, and to me, that tiny patch of grass seemed like a farm.

When I moved to Tennessee years later, I saw real farms for the first time—huge fields stretching as far as I could see. It was nothing like the concrete and buildings I grew up with. But now I watch sadly as these farms are sold off to build new neighborhoods and subdivisions.

This isn't just happening in Tennessee—it's happening all across America. But losing farmland is just one part of a bigger problem: *we've lost something even more important—the knowledge of how to grow our own food.* Most people today believe they can only get food from grocery stores and big companies. We've forgotten that for thousands of years, people knew how to feed themselves by working with nature.

This hits close to home for me. I have four kids, and I see more and more of my friends' children getting sick from their food. Many can't eat normal things like wheat or milk. In fact, most people would say they have a food allergy of some kind. Yet these are foods that humans have consumed for thousands of years. How could this be? Well, it's not because these foods are naturally bad. It's because the companies that make our food use unnatural processes, and add all sorts of chemicals—chemicals that many other countries don't even allow in their food.

My own journey with health and fitness began at age 13, driven by a simple desire: I wanted to build a body that wouldn't age, that would have more energy than my future kids. This passion led me deep into the world of fitness, nutrition and supplementation. But what I discovered was disturbing. The very supplements marketed as "healthy" were often more processed and chemical-laden than the foods they were meant to replace. The fitness industry, like so many others, had been corrupted by the pursuit of profit over health.

Now don't get me wrong, I'm a capitalist. I believe that businesses must make a profit, but that profit must come from activities that help humanity, not hurt it. I've spent years building and working with companies that make products. Many of those products are food and nutritional products. I've seen behind the scenes how these businesses work. I've watched good companies turn bad after bigger corporations buy them. I've seen how they use tricks and loopholes to call things "organic" when they're not really healthy at all.

We've been fed a fat lie. We're told that healthy, natural food has to be expensive. We're told we can only get good food from fancy stores or farmers' markets that cost way too much and are often inaccessible. Even worse, we're told we can't grow food ourselves—that only big companies can do that for us. Don't believe me? Watch any 30 minute TV show and observe the commercials. They're all "food" commercials—mostly heavily processed food—and drug commercials. Why is that? One makes you sick and the other profits off of your sickness.

The amazing thing is that the answer to breaking the cycle, is a simple truth. Everything we need to be healthy in life is freely available to us. It's part of God's beautiful design for this world.

That's why this book by Jabez is so important. He shows us something that big food companies don't want us to know: healthy food shouldn't cost a lot of money. Healthy food isn't inaccessible. *Nature gives us everything we need, for free.* We've just forgotten how to get it.

Think about this: for thousands of years, people knew how to grow their own food. It wasn't special knowledge—it was something everyone learned, like learning to read or write. Today, we're at risk of forgetting these important skills completely. We can't let that happen. *Being truly independent starts with knowing how to grow your own food.*

FOREWORD

The Naturalite Lifestyle that Jabez introduces isn't just another diet. While other diets focus on what you should and shouldn't eat, Naturalite focuses on something more important: where your food comes from and how it's made. This is what makes this book so special—it's now about changing what you eat, but about reconnecting with the natural ways people have grown food for thousands of years.

This book is more than just a guide to gardening. *It's a map to food freedom*, a blueprint for breaking free from depending on big corporations and governments for your food. Through his own stories and practical wisdom, Jabez shows us that eating healthy doesn't require lots of money—it just requires understanding how to work with nature.

We live in a time when we have more abundance and access than ever before. So why are so many people sick? Something is clearly wrong with our food system. But as this book shows, the solution is growing all around us. We just need to learn how to see it.

This connects deeply with my own mission of building businesses that make the world a better place. Through my company Metacake, I've helped hundreds of businesses grow. But what excites me most about this book is how it can help change not just individual health, but entire communities through the power of home agriculture.

This book isn't just important—it's urgent. We can't afford to lose these skills.

Whether you're a parent worried about your kids' health, someone that's busy running a business, or just tired of paying too much for "healthy" food that isn't really healthy, you'll find valuable wisdom in these pages.

Read this book carefully. Share it with others. The future health of humanity depends on more people understanding and using these ideas.

Kenneth Ott

Co-founder, Naturalite
Co-Founder, Metacake
Author, "The Ecommerce Formula"

For more information about me, visit: kenott.com

INTRODUCTION

THE HIDDEN COST OF CONVENIENCE

AMERICA HAS MASTERED both good and evil. Living here without awareness of how things operate can lead to serious problems, but having awareness can open up incredible opportunities. Right now, I believe there's a tremendous opportunity for anyone willing to learn the practice of agriculture—how to grow food. At the same time, it's becoming increasingly risky not to understand how food is grown naturally. Just look around: grocery stores and restaurants are overflowing with processed, unhealthy foods.

One of the most significant consequences of consuming these foods is weight gain, and it's no coincidence that nearly 80% of Americans are now overweight.

I became a best-selling author by writing *Life Matters So Let's Eat Like It!* In that book, I shared how I lost 100 pounds while still enjoying the foods I love. The key wasn't giving up my favorite meals. It was learning how to find more natural versions of the foods I already enjoyed. By making simple switches, I unlocked a way of eating that helped me lose weight while still enjoying what I loved. Since then, I've helped thousands of people lose weight through my books, coaching, our Naturalite Weight Loss Community and courses like the *Eat What You Love and Weigh Less Reprogram.*

Along this journey, I realized something powerful: there was no category of food consumption that focused on cutting out heavily processed foods and chemicals. So, I created one. I'm not a vegan. I'm a **Naturalite**. Being a Naturalite is a lifestyle choice. It's about reading the ingredients, choosing natural food options, and cutting out the unnatural processes and chemical additives that contribute to weight gain and other concerning side effects. But it goes beyond that. It's also about reconnecting with the natural process of growing our own food, so we're not entirely dependent on grocery stores and restaurants to provide for us.

We live in a society that prioritizes convenience over connection. Grocery stores and restaurants have made it easy for us to disconnect from what we eat, how it's made, and where it comes from. This disconnection has left many of us unaware of how our food choices impact our weight and habits, relying on a system that doesn't prioritize our needs. My mission with this book is to empower you to take back control. Whether you have a backyard garden or just a few pots on a balcony, growing your own food can transform your relationship with it. My goal is to show you how simple and rewarding it is to start this journey, even with little to no money. By embracing the Naturalite Lifestyle, you'll gain the tools to eat better and live in harmony with nature.

This book isn't just about food…it's about freedom. It's about breaking free from processed foods and convenience culture to embrace a life where you're connected to your food and the natural world. Together, we can cultivate a future where natural food isn't the exception but the standard! Let me be clear, growing food takes intention and consistency. And just so you know, I started my first garden only a few years ago. This is a lifelong journey for me, and I'm learning as I grow quite literally. It was imperative that I understood and applied the principles I share in this book. I made the effort to work intentionally, and my success in taking a plant from seed, seedling, or propagation to producing food has skyrocketed. If I can do it, so can you.

I'm writing this book because I believe that if we want the food industry to change, more of us need to start practicing agriculture and understanding how natural food is produced. This journey has been an eye-opener, lifting the veil on what's truly natural and what's not. My hope is that this book gives you the tools and knowledge to grow your own food 100% naturally, empowering you to make better choices for yourself and your family. Choices that can protect you from weight issues and the negative consequences of relying on unnatural food sources.

To truly be free, we must have control of food production. That freedom begins with understanding the earth and the natural elements that provide for us abundantly at no cost. My agricultural journey began in 2020, during the pandemic. I started small, growing garlic, rosemary, basil, oregano, and even attempting an orange tree in a pot. Later, I planted that orange tree in the ground, but it didn't survive. It was a tough lesson that taught me a lot, and one I'll share more about later in this book. What has fascinated me most on this journey came about three years in, around 2023, when I had a life-changing realization: The more I GROW, the more I see EVERYTHING we need is FREE!

CHAPTER 1

THE FIRST STEP

IF IT WEREN'T FOR THE PANDEMIC, I don't think I'd even be writing this book. 2020 and 2021 were, without a doubt, the most interesting, nerve-racking, and life-changing years of my life. When everything shut down, I made a decision: I was going to use the isolation and extra time to make some big changes. In just the first four months, I grew more as a person than I had in the previous four years. I dove headfirst into self-development and took real action. One of the first things my wife and I did was remove the TV from our bedroom. I also made a choice to stop listening to rap music and replaced it with something better. I told myself: no more pointless content. I'm only consuming things that will make me better.

From the moment I woke up to the moment I went to sleep, I was on a mission for knowledge that I could apply to put my family in a better position. I watched educational YouTube videos, listened to podcasts, and read books. The most valuable book I read during that time was the Bible. I didn't read it like I had before. This time, I read it from a different perspective. An empowering one. I searched for wisdom and principles I could apply to my character and my life, with one goal... to see if it would make me a better man.

I could probably write a whole book about this period of my life, but to sum it up, my mind started going through some serious changes. I was evolving **big time**! One day, during this time, I drove to the grocery store in Miami to pick up some essentials for the house. When I walked into Publix, I froze. The place looked worse than when a hurricane was coming. Shelves were empty. No produce, no chicken, no steak. Nothing. Even the fish section was wiped out. I'd never seen anything like it. I called my wife, and she turned on the news. They were talking about a food shortage, and to top it off, you could forget about finding toilet paper! I left Publix and headed to Walmart, hoping for better luck. Same deal. Walmart was full of empty shelves. I tried another Publix, but it was the same thing again. I grabbed what little I could and drove home.

That night, something hit me deep. I felt uneasy. **Really uneasy**. It dawned on me that I hadn't been thinking about how to provide food for my family beyond the grocery store. We were completely dependent on a system that could fall apart at any moment. That night, I sat down and prayed, asking for forgiveness. I realized I had been overlooking something deeply important... taking responsibility for ensuring my family would be okay if things got worse. That moment changed me. It was a wake-up call I couldn't ignore, and from that point forward, I knew I had to step up.

The old me would've blamed the system, but the new me was ready to take responsibility. That night, I asked myself a simple but life-changing question: *What can I do to get food from nature, not just the store?* At the time, we were living in a townhouse with a tiny backyard, so starting a full garden wasn't an option. That realization scared me a little, but I wasn't going to let fear stop me. Besides, I didn't know anything about growing food anyway. I had never grown an edible plant from start to finish in my life. But I was determined to try and **stay consistent!**

I had to start somewhere, so I made a plan. My first step was to grow herbs in pots, both inside and outside our home. I figured it would give me a feel for growing and caring for plants without being too overwhelming. At the same time, I decided to prepare for other possibilities. I bought a pellet gun. There are plenty of ducks in Miami, and they're easy to get close to. As a backup, I also stocked up on big bags of rice. If things got worse, that combination would at least give my family a little more time.

That night, I prayed hard, asking Yahweh for two things: time and guidance. I knew it might take few years, maybe more, to learn how to grow enough food to support my family. At the time, I had no following on social media and hadn't even written my first book, *Life Matters So Let's Eat Like It.* Money was tight, and I was painfully aware that if the grocery stores really ran out of food, we'd be in serious trouble. So, I had a backup plan: if it came to that, we'd pack up and move in with family out in the country.

I'll be honest, I felt like I hadn't been handling my responsibilities as a man. If I didn't feel safe, how could my family feel safe? That prayer is etched in my memory. I meant every word when I asked the Creator to give me time to learn what I needed to learn. I made a promise that night, a promise I refuse to break. *I will act on what I learn, and I will not waste the time I am given!* I was determined to learn how to grow food and find a way to make money online so I could create income from anywhere, no matter the circumstances. Looking back, I see now that something deeper was happening. Psychologically, I had reached a new level of awareness about my role as a man and the responsibility I have to take care of my family no matter what.

Right after that prayer, I didn't waste any time. I jumped on YouTube and started watching "how-to" videos. And that brings me to the first important lesson I want to share with you, or should I say, the first *seed* **I want to plant in your mind**... Start now and use the free information that's available. You can learn how to grow just about anything using the resources that are freely accessible online. To illustrate this point, let me tell you what happened next. By the end of 2021, I went viral on TikTok, and my social media following started growing fast. At the time of writing this, I have over 800,000 subscribers on YouTube, 600,000 followers on Instagram, and more than 500,000 followers on TikTok. I even started posting on Facebook and quickly passed 380,000 followers there.

Throughout this entire process, I loved engaging with people in the comments of my posts. Over time, I started noticing something. People were asking me questions about how to grow certain plants. Questions they could easily find answers to with a simple YouTube or Google search.

Here's the Great News!

We are blessed to live in a time where information is at our fingertips. All you have to do is know where to look and what to ask. After a couple of years of gardening, I've learned that the most crucial skill in learning how to grow plants successfully is knowing where to find the step-by-step answers you need to grow the things you want to grow in your area. Let me be clear, this book isn't about giving you specific growing instructions for each plant. Instead, it's about building your awareness of foundational principles and showing you how to find the specific information you'll need to start your own garden and successfully grow your plants through to harvest.

I've intentionally designed this as a book of principles, not a book of details. The details of agriculture have been documented for centuries, and there's an abundance of free information available online. Once I learned to ask the right questions in the right places, I have been able to find all the specifics I need to succeed in growing anything I set my mind to. My main resources are YouTube, ChatGPT, and Google, but I also tap into social media platforms for insights and connections. Below, I'll give you examples of how I use each of these tools and, more importantly, how I frame my questions to get the specific, step-by-step guidance I need.

It's worth mentioning that while YouTube, ChatGPT, and Google are my go-to platforms as of now, the future will undoubtedly bring new and potentially even more powerful tools. I encourage you to stay open to innovation and keep an eye out for emerging platforms. This awareness/mindset doesn't just apply to agriculture, it applies to every area of life. We are living in the information age, and there are incredible opportunities for those who choose to act on the knowledge that's freely available to us.

Now, let me walk you through how I use these platforms and the way I structure my questions to get clear, actionable advice for growing what I want. YouTube has probably been my most-used resource because I'm a visual learner. It's an incredible platform that provides visual, hands-on guidance, making it especially helpful if you like seeing each step in action. The secret to getting the most out of YouTube is structuring your search with as much detail as possible, focusing on the plant you want to grow and the specific environment you're working with.

THE FIRST STEP

For example:

- Want to grow peppers indoors? Search: "How to grow peppers in pots inside an apartment."

- Working with a backyard in a specific area? Try: "Setting up a raised bed to grow vegetables outside in Houston, Texas."

By being specific about what you're growing and where, you can find videos that address your exact needs. These step-by-step guides make it easy to follow along and put what you've learned into action. I'm still amazed at how much valuable, practical information is available for free on YouTube and how effective it can be when applied!

If you're not familiar with ChatGPT, it's an artificial intelligence program that you can access online or through an app on your phone. There's both a free version and a paid version. The free version is incredibly useful, and the paid version takes it to the next level. Think of ChatGPT as a knowledgeable friend who's always ready to guide you through the growing process, one question at a time.

The key to using ChatGPT effectively is to ask clear, specific questions to get actionable answers. For example:

- "What steps do I need to follow to plant tomatoes directly in the ground in Florida?"

- "Can you guide me on planting tomatoes in the ground in clay soil during springtime in North Carolina?"

The more detailed your question, the better ChatGPT can tailor its guidance to your needs. To get the most useful answers, it's crucial to ask for step-by-step instructions and frame your question with the expertise you'd like ChatGPT to emulate. For example:

- "Can you guide me on planting tomatoes directly in the ground in Florida? Please provide a step-by-step explanation, starting with preparing the soil, choosing the right tomato variety, and planting techniques, and ending with tips for watering and maintenance. Answer from the perspective of someone with over 30 years of gardening experience who has successfully grown thousands of pounds of tomatoes. Make the explanation simple enough for a fifth grader to follow and achieve great results."

This framing is powerful because it allows ChatGPT to respond as if the advice is coming from a highly qualified expert. By prompting ChatGPT to answer from the perspective of someone with decades of experience, you ensure the response is practical, detailed, and designed to maximize your success rate. It transforms ChatGPT from just another tool into a virtual mentor who can guide you step-by-step with confidence and clarity. The answers you'll get from posing questions like this are mind-blowing. By tailoring your prompts this way, ChatGPT quickly becomes one of the most powerful tools for gathering valuable information, whether you're growing a small herb garden, managing a backyard vegetable patch, or starting a 20-acre farm.

Google is another powerful tool, especially for finding resources and connecting with others in your area who are into agriculture. I often use Google to search for local nurseries. I love visiting nurseries and talking with the people there, especially the owners. They have a wealth of knowledge about growing plants and know what grows best in your area based on their experience. Google acts as a direct map to finding these people.

Another resource that has helped me tremendously is social media. Platforms like Instagram and Facebook are excellent for connecting with the agricultural community and learning from others. On Instagram, I often search phrases like "gardening in Miami" or "growing food in Miami." This pulls up posts from people who are actively growing food in my area. If I find a page that catches my attention, I'll check out their work and often reach out to them directly through a DM to connect.

One thing I've learned is that the gardening and growing community is one of the most welcoming groups you'll ever find. Most of the time, all it takes is letting people know you're interested in connecting, and they're happy to share their knowledge, tips, and resources.

The Groups feature on Facebook is an extremely powerful tool. These groups bring together people with shared interests like gardening, agriculture, or growing food. All you have to do is search for groups in your area and join the ones that match your goals. Inside these groups, you'll find a wealth of information, support, and resources, often from people who live in your region and understand your growing conditions. The amount of help and insight available in these communities is truly amazing!

Take Action!

Once you have the information you need, the next step is to **take immediate action**. Whether you've gathered instructions from YouTube, Google, or someone you connected with on social media, don't wait and get started! Begin by gathering the materials you need and follow each step as instructed. Before we move on, I want to plant this seed in your mind: The most valuable thing for anyone looking to grow food is understanding how to use the information resources available on your phone. This will allow you to get the specific details you'll need to successfully grow each plant you choose. Trust me, mastering these tools has been the cornerstone of my agricultural journey.

But no matter how many resources you use, there's no better teacher than practice. I say this from experience because I'm still learning as I GROW! If you're wondering how much time it takes to tend to your plants, the good news is I only spend on average about 10 minutes a day on my two gardens and fruit trees. On days when it rains, I usually don't spend any time at all. Once you find your rhythm, it doesn't take as much time as you might think to care for plants and grow them successfully.

Now, let's move on to the next chapter, where we'll explore a topic that fascinates me deeply: understanding your environment and what plants flourish there.

CHAPTER 2

UNDERSTANDING YOUR ENVIRONMENT: GROWING ZONES

WHEN I FIRST STARTED GROWING FOOD in 2020, one of the plants I decided to grow was an avocado tree. I love avocados, so I thought, "Why not grow my own?" One day, I took the pit out of a Hass avocado I bought from the grocery store. I had heard you could germinate (which is a method of placing a seed in moisture to cause the roots to come out of the seed) the seed at home, so I searched YouTube and found a video explaining the process. The instructions were simple: stick three toothpicks into the avocado pit, place it halfway into a glass of water, and let it sit until roots began to grow. Then, when the baby tree sprouted out of the top, it would be ready to plant in soil.

I was excited and started the process right away but let me tell you… it took much longer than I expected. Day after day, I'd check the seed, waiting for roots to appear. Finally, after what felt like forever, the roots grew, and soon enough, a tiny tree began to emerge. I was super excited! Once it was big enough, I transplanted the seedling into a small pot with soil, and it continued to grow. I can't remember exactly how long it stayed in that pot, but eventually, I decided to move it outside. I imagined it growing tall, full of avocados, right in my backyard. But then something happened. After a while, my little avocado tree started to look bad. The leaves wilted, the color faded, and it just didn't look healthy. I was confused and discouraged. I had put so much time into caring for this tree, and I couldn't figure out what went wrong.

I started researching again, watching more YouTube videos on how to heal a sick avocado tree. But something didn't sit right.

The videos I found were from people who lived in completely different places such as California, Texas, even as far as Australia. And then it hit me. Maybe this type of Hass avocado doesn't grow well in Miami's climate. I looked it up, and sure enough, I learned that tropical climates like Miami are better suited for tropical avocados, not this Hass variety I was trying to grow. Tropical avocados are bigger and thrive in the heat and humidity of South Florida.

My tree didn't survive, but the experience planted something even more important: a seed of knowledge. This experience taught me how vital it is to understand what plants grow best in your specific area. This awareness set me on a path of discovery that I'm still on to this day.

Understanding Growing Zones

Growing zones, also known as USDA Hardiness Zones, are regions that have been defined to help gardeners and farmers identify what plants are most likely to thrive in their specific area. These zones are determined based on the average lowest winter temperatures in a region, providing a guide to what plants can survive and grow successfully in different climates.

The United States is divided into thirteen growing zones: Zone 1 is the coldest (think Alaska), and Zone 13 is the warmest (like Puerto Rico). Each zone is further divided into sub-zones, such as 8a or 8b, which reflect smaller temperature ranges within a region. These zones are an essential tool for anyone looking to grow food because they take the guesswork out of choosing plants that will thrive in your environment.

Why Growing Zones Matter

If you've never thought about growing zones, consider this... It's like trying to grow a cactus in a swamp. It just doesn't work! Understanding your zone helps you avoid wasting time and energy on plants that aren't compatible with your climate.

For example:

- **Zone 10**, which includes Miami, is ideal for tropical plants like mangoes, papayas, and tropical avocados.

- **Zone 5**, found in places like Wisconsin, is perfect for hardy crops like apples, cherries, and root vegetables.

Growing Zones Across America

Here's a quick breakdown of the zones in the United States:

- **Zones 1–3**: Alaska and northern areas (extremely cold climates)

- **Zones 4–6**: Midwest and Northeast (cold winters, mild summers)

- **Zones 7–8**: Southern states and parts of the West (mild winters, warm summers)

- **Zones 9–11**: Southern California, Florida, and Hawaii (warm to tropical climates)

How to Use Growing Zones to Your Advantage

Once you know your growing zone, it's easier to figure out what to plant. For a detailed list of plants suited to your zone, I recommend asking ChatGPT, Google and YouTube for ideas or talk to experienced growers in your area. I've learned a lot from people I connect with on social media and at local nurseries. They're often happy to share what crops work best for them and why. By understanding your zone and choosing plants accordingly, you'll set yourself up for gardening success and enjoy thriving fruits, vegetables, and herbs that are perfectly suited to your environment.

Learning from Other Countries

One fascinating thing about growing zones is that they're not just limited to the U.S. Similar climates exist all over the world, and that means plants from other countries can thrive in your area if the conditions match.

For example:

- **Zone 10 (Miami):** Plants like soursop from the Caribbean or dragon fruit from Southeast Asia grow wonderfully in Miami's climate.

- **Zone 7 (Georgia):** Crops like figs from the Mediterranean or persimmons from East Asia thrive there as well.

By looking at what grows well in similar climates around the world, you can expand your options and grow unique plants that aren't as common in your region.

The Road Trip Lesson

I've had the privilege of driving across the United States, from Washington state to Wisconsin, from Miami to Delaware, and beyond. Every road trip reminds me of how the landscape and the plants that thrive in it changes as you travel. What grows naturally in one area can be completely different just a few hours down the road. Nature gives us clear clues about what grows best in each environment. In Miami, you'll find lush tropical plants, while in Georgia, you're surrounded by towering pine trees and vibrant red clay. The more you observe, the more you realize that everything in nature has its place and its perfect conditions. The next time you travel, take a second to pay attention to your surroundings. Notice the plants, the trees and even the climate in the different destinations. This will help you once you start growing your own.

Understanding your growing zone is one of the most valuable things you can do as a grower. It saves you time, money, and frustration by helping you work with nature rather than against it. If you love a particular fruit or vegetable that doesn't grow in your zone, don't worry. Often, there are similar plants that thrive in your area and still provide the flavor or function you're looking for. For example, if cherries won't grow in your warmer climate, mulberries might be a perfect alternative with a comparable flavor and better adaptability to heat.

The key is matching what you want to grow with what your environment naturally supports. Once you understand this principle, the possibilities are endless. By paying attention to your environment and choosing plants suited to your zone, you set yourself up for success. Nature has already done the hard work of determining what grows best where. It's up to us to observe, listen, and learn.

Let this awareness serve as a foundation for your growing journey. The more you grow, the more you'll see that everything you need is already within your reach!

CHAPTER 3

WATER AND SUNLIGHT: LIFE'S ESSENTIALS

AGRICULTURE HAS BEEN OVERCOMPLICATED for far too long. At its core, there are just four foundational principles: **water, sunlight, climate, and soil**. When you understand these principles and combine them with knowing where to find the specific details for the growing process of what you want to grow, success becomes natural. You don't need to be an expert to get started. As long as you are willing to care for your crops and pay attention to what they need, you can grow just about anything!

In this chapter, we'll explore two of these foundational principles... **water and sunlight**. I'll explain how the awareness of these principles are helping me have success and how they can help you just the same.

The Mint Plant that Taught Me About Water

When I first decided to start growing plants, one of my first experiments was with a mint plant. I bought it already grown from the store, imagining how nice it would be to have fresh mint for drinks. My plan was simple... let the plant grow bigger and stronger before picking from it so it would keep producing.

At first, I watered it every couple of days, thinking I was doing everything right. But after a couple weeks, the plant hadn't grown at all. In fact, it looked worse than when I bought it. The leaves were drooping, and the whole plant seemed like it was struggling to survive. I started questioning what I was doing wrong. Was I watering it too much? Was it not getting enough sunlight? To test my theory, I moved the plant to a sunny spot by the window and waited a few days before watering it again. But still, nothing changed.

One day, as I was about to fill my watering can from the kitchen sink, a thought hit me. *Could it be the tap water that's killing this plant?* I reflected on other plants I'd tried to grow in the past that also failed, and it clicked! I had been using tap water on all of them. At this point in my life, I had already stopped drinking tap water because of the chemicals like fluoride and chlorine used to treat it. I figured, if it's not good for me, it probably isn't good for my plants either.

So, I decided to switch to spring water. And soon after, the leaves perked up, stood tall, and the mint plant made a full recovery. Even better, it started growing new leaves. That was my first big lesson in agriculture. Tap water, especially for small plants, isn't ideal. Switching to natural water sources can make all the difference.

The Reservoir That Sparked an Idea

As I started growing more plants, buying spring water from the store became expensive. I needed a better solution. That's when a memory from a road trip with my brothers came back to me. We were driving through the desert in Southern California when we spotted a reservoir nestled in the hills. Despite being in the middle of a dry desert, the area around the water was lush and green. I remembered talking about how water brings life to everything around it. That memory sparked an idea. *What if I used lake water for my plants?*

I grabbed a few 5-gallon buckets and headed to the nearest lake to test the idea. The results were incredible. My plants thrived, just as they had with spring water. But as my garden grew, so did my need for water. Getting buckets from the lake eventually turned into a time-consuming mission.

The Power of Rainwater

One of the most fascinating things I've observed about rainwater is how quickly it brings life back to the land. During dry spells, everything looks brown and lifeless. The grass becomes brittle, the trees seem dull, and plants struggle to stay upright. Then, after a heavy rain, the transformation is almost immediate. The grass turns vibrant green, flowers bloom overnight, and everything comes alive again. It's as if nature had been holding its breath, waiting for that rain to fall.

This made me think about the power of rainwater compared to other water sources. Rain isn't just water. It's natural, unfiltered, and filled with nutrients absorbed from the atmosphere. Unlike tap water, which is treated with chemicals like chlorine, rainwater is pure. Watching how quickly plants responded to rain made me wonder what would happen if I used rainwater exclusively for my garden.

Around the same time, I remember a day when the sky turned dark, the wind picked up, and I could tell a storm was coming. As the rain started pouring, an idea hit me… *Why not catch the rainwater?* I ran outside and placed the same buckets I had been using to get lake water under the spots where water was streaming off the roof. In no time, the buckets were full. To collect even more, I bought a large plastic trash can to store water for dry spells. To check out a video I did showing this process, simply open your phone's camera, point it at the QR code, and tap the link that appears on your screen

The Role of Sunlight

Just as water is essential to the growth of your crop, so is sunlight. But not all plants need the same amount of sunlight, and their needs can change as they grow. When I first started growing baby fruit trees, I made the mistake of putting them in direct sunlight right away. The intense sunlight burned their leaves, and the trees eventually died. It wasn't until I tried growing the same trees in a shaded area that I noticed a difference. In indirect sunlight, the trees grew stronger and healthier. Once they reached about 20 inches tall, they were able to thrive in more direct sunlight.

WATER AND SUNLIGHT

This experience taught me an important lesson: smaller plants often need indirect sunlight to grow strong enough before they can handle direct sunlight. For other plants, like vegetables, sunlight requirements vary. Some vegetables need full sun all day, while others thrive in partial shade. Paying attention to your plant's specific sunlight needs is critical at every stage of growth. Make it a habit to check the information resources we went over earlier in this book to learn your plants' sunlight requirements as you GROW. Understanding these needs can make the difference between a struggling garden and one that thrives.

Water and sunlight are the foundation of life. Without them, no plant can grow. But it's not just about having water or sunlight. It's about providing the right kind of water and the right amount of sunlight for your plants. As I continued to grow, I realized that water, sunlight, climate, and soil work together in perfect harmony. Each one plays a critical role, and when you understand how they interact, you can grow almost anything. Water and sunlight remind me of life's simplicity: everything we need is already around us. It's up to us to pay attention, learn, and work with the natural order of the world we live in.

CHAPTER 4

SOIL – THE FOUNDATION OF GROWTH

THIS CHAPTER IS ESPECIALLY IMPORTANT to me because I believe it may be the most critical one in this entire book. The number 4 holds deep significance. It's a foundation number. Four corners create stability, and the number 4 also represents the earth. We've all heard of the four directions: North, South, East, and West, and the "four corners of the earth." Just as the number 4 symbolizes a strong foundation, soil is the foundation for growing the things that support us freely.

When I imagine the future, I see thriving gardens bursting with nutrient-dense food. I see vibrant herbs, fruits, and vegetables flourishing in soil that's rich, dark, and alive. But to achieve this vision, we need to confront a major challenge: **our soil is losing its nutrients at an alarming rate.** If we want to leave a better future for our children's children, we must address the urgent problem facing our planet's topsoil.

Ever since I was young, soil has fascinated me. I loved the way it felt in my hands and the earthy smell that came from it. I lost count of the times I came home covered in soil and got scolded for being dirty. Looking back, I realize there was something special about soil that my spirit recognized, even as a child. I remember noticing the little creatures crawling around in it and how soil would have differences at each one of my friend's houses. I noticed differences in. texture, color and moister levels. I also noticed times we would dig into the ground with a shovel, some soil was easy to dig while other soil was hard and we would need a pickaxe.

The soil I loved most was soft, rich, and dark... almost black. I noticed that plants seemed to grow all over this kind of soil, even though I didn't understand why at the time. Years later, when I started growing plants in 2020, I began to connect the dots.

The Orange Tree Lesson

Remember earlier in the book when I mentioned the orange tree we started growing in 2020? Let me tell you what happened. We took seeds from an orange we bought at the grocery store, and I had my children germinate them. When the seeds sprouted, we planted them together. I actually recorded when we did it and posted it on Instagram. To check out the video I did showing this process, simply open your phone's camera, point it at the QR code, and tap the link that appears on your screen.

Out of the four seeds, only one grew well in the pot; the other three didn't make it. The tree that survived we eventually named Orangey.

We grew Orangey in a pot for about a year and a half. Then, when we moved, I had a new idea. This was around the time I decided to become Jabez Mango Seeds and set a goal to bring over one million fruit trees to South Florida. A goal that eventually evolved to bringing over one million fruit trees to America. My initial plan was simple. I wanted to plant fruit trees in areas where they could thrive on their own, like near lakes or roadsides, where people could come by and pick fruit whenever they wanted.

Orangey was one of the first trees we planted as part of this vision. We found a spot off the road in Homestead, Florida, near a lake. I thought it was a good spot at the time. I even recorded the process of me and my son planting it in the ground. To check out the video I did showing this process, simply open your phone's camera, point it at the QR code, and tap the link that appears on your screen.

SOIL

When we planted Orangey, I noticed the ground was incredibly hard, almost like dry clay and rock. I had to use a pickaxe to break through the ground just to make a hole for the tree. For almost two years, Orangey grew... very slowly. We'd visit every so often to water it and check on its progress.

But then a few months passed without us visiting. When we finally went back to check, the tree was dead. This experience taught me a critical lesson: the soil conditions where we planted Orangey were not right for the tree to thrive. The ground was too hard, lacking the nutrients and structure that young roots need to grow strong. I also believe the seed itself, coming from a store-bought orange, wasn't ideal. (I'll be sharing some advice on seeds in the next chapter)

The bigger takeaway, though, was this... while my idea of planting fruit trees in public spaces was well-intentioned, it wasn't practical for ensuring the trees' success. I realized that planting fruit trees on people's properties was a better approach. On private land, we could control the soil conditions, provide better care, and give the trees the attention they needed to survive and eventually bear fruit.

Orangey's story is a reminder that good soil is the foundation for growth. Without it, even the best intentions and most careful efforts may not lead to success. Soil matters. It's the difference between a struggling tree and one that thrives and bears fruit for years to come.

My First Garden Lesson

Another moment that truly opened my eyes to the importance of good soil happened with my first tomato plant. I saw a video on YouTube where someone took a slice of tomato, buried it in soil, and grew a plant from the seeds in the slice. And it worked! One slice of tomato grew multiple plants, and those plants produced tomatoes. I was inspired and decided to try it myself. I bought some soil from the store, filled a small plastic bag, placed the tomato slice on top, and covered it with a little more soil. I kept the soil moist by watering it regularly, and sure enough, the plants started to grow! Excited by my success, I decided to take it a step further. I dug a small garden in the ground next to my house.

I transplanted the tomato plants into the ground, along with a bell pepper plant and some kale. But then something unexpected happened. While the plants grew, they didn't thrive. My tomato plant gave me a few small tomatoes before it died, and the other plants didn't do much better. That experience taught me another important lesson. That the soil in my yard wasn't good enough. It lacked the nutrients that plants need to grow strong and healthy. This realization sparked a journey of learning for me. I started watching more videos, asking experienced gardeners for advice, and studying soil nutrients.

That first attempt may not have given me a thriving garden, but it gave me something even more valuable: the understanding that healthy soil is the foundation of a successful garden. Without it, plants will always struggle, no matter how much effort you put into them.

I think at this point it's important to note I've come to realize that successfully growing plants starts with having the right mindset. It's a journey that requires curiosity, observation, and commitment. My mindset is simple. Every problem that comes up has a solution. I just have to find it. I'm not giving up on my plants because I care about them! If other people can grow plants successfully, so can I. All I need to do is learn, apply what I've learned, and keep adjusting as I go.

If you'd like to grow food successfully, I encourage you to take on this mindset. It makes the journey more rewarding and guarantees results when you stay curious, committed, and willing to learn.

The Hidden Crisis Beneath Our Feet

As I reflected on my gardening experiences, I realized how much success or failure depends on the soil beneath us. Soil isn't just dirt; it's the foundation of all plant growth. It holds the nutrients that crops, fruits, vegetables, herbs, and trees need to thrive.

But here's the problem: the quality of our soil is declining both in the U.S. and across the world. This isn't just a gardening issue; it's a global crisis affecting the food we eat, the environment we live in, and even the economy.

Let me share some key points I've learned about this hidden crisis:

1. Declining Soil Nutrients

Since 1950, U.S. soils have lost a significant amount of nutrients. For example, there are studies showing levels have dropped by 42%. This impacts how well plants grow and reduces the nutritional value of the food they produce.

2. Soil Fertility Around the World

Different regions face different challenges. In Africa, because of mass farming methods that are not in alignment with nature, soil nutrients are being depleted faster than they're replenished, making farming increasingly difficult. On the other hand, some parts of Asia have improved their soil nutrients by adopting better agricultural practices.

3. The Problem of Soil Erosion

Erosion (The wearing away of soil) caused by farming methods is a major issue in places like the U.S. Midwest. Over the last 160 years, about 57 billion tons of topsoil have been lost in this region. Since most nutrients are in the topsoil, this loss forces farmers to rely on fertilizers, which are costly, can harm the environment and continues to make the soil worse.

4. Economic Consequences

Poor soil doesn't just hurt plants; it hurts farmers' wallets. For example, U.S. corn farmers spend more than half a billion dollars every year on extra fertilizers to make up for poor soil.

5. Global Impact

According to Wikipedia, about 75 billion tons of soil are lost annually due to erosion worldwide. This damages farmlands and costs hundreds of billions of dollars each year, showing just how urgent it is to restore and protect our soil.

Understanding these issues is critical if we want to grow healthy, nutrient-dense food. Restoring soil health isn't just about solving problems for today it's about creating a better future for the next generations.

The Simplicity of Composting

One way to start healing our soil is through composting. Composting isn't as complicated as it might seem. At its core, it's about combining even parts of three simple ingredients:

1. 1/3 **Soil from the Ground** – Speeds up compost turning into usable soil.

2. 1/3 **Carbon** – Brown, dry materials like dead leaves, wood chips, or straw.

3. 1/3 **Nitrogen** – Fresh materials like grass clippings, fruit scraps, and vegetable peels.

To start, layer these ingredients evenly, keep the mix moist, and let time work its magic. Over weeks and months, the materials will break down into rich, dark soil full of nutrients that your plants will love. The easiest way I found to start composting is by using the garbage bin method. It's simple, effective, and perfect for beginners. To check out a video I did showing this process step-by-step, simply open your phone's camera, point it at the QR code, and tap the link that appears on your screen.

The Purple Sweet Potato Lesson

One of my most memorable gardening experiences came from a purple sweet potato plant that started as a sprout in my compost pile. It grew from some purple sweet potato skins I had thrown into the compost about a month earlier. I had no idea that the skins could actually produce another plant, so when I saw the sprout, I didn't recognize what it was. Even though I wasn't sure what it was, I decided to plant it in my garden.

To my surprise, the plant took off! The vines spread everywhere. Me being curious, I posted a video of the plant on social media to see if anyone could identify it. People quickly responded, telling me it was sweet potato. Then I remembered the purple sweet potato peels that I threw in the compost a couple weeks prior and looked up pictures of purple sweet potato leaves. Sure enough, they were identical! I even discovered that the leaves were edible and tasted like spinach.

When it was time to harvest, I was both surprised and a little let down to find only one purple sweet potato in the soil. After pulling up the vines, I noticed something else. The soil where the plant had grown looked dry and lifeless. To check out a video I made about this experience, simply open your phone's camera, point it at the QR code, and tap the link that appears on your screen.

That's when it hit me! Plants pull nutrients out of the soil as they grow, and if I don't replenish those nutrients, the soil becomes depleted.

A Natural Cycle

This experience made me reflect on how nature naturally replenishes nutrients. In my garden, I noticed something fascinating: the soil near my avocado tree and comfrey plant was thriving. The tree's fallen leaves and the organic matter from the comfrey plant fell to the ground, slowly breaking down and returning nutrients to the soil. Meanwhile, the surrounding plants continued to grow stronger.

This is a perfect example of a growing method I'm super fascinated with called permaculture. Permaculture mimics nature by creating a self-sustaining cycle. Plants feed the soil, and the soil feeds the plants. I encourage you to check out this video I like that explains permaculture. I encourage you to check out this video I like that explains permaculture. Simply open your phone's camera, point it at the QR code, and tap the link that appears on your screen to watch the video now.

When we work *with* nature instead of exploiting it, we no longer need to rely on synthetic fertilizers or disruptive farming practices. Instead, the system becomes balanced and self-sustaining.

Another fascinating connection I've become aware of is how the Earth's surface is like the skin on our bodies. When our skin is cut or broken, our body immediately begins working to cover the wound, regenerating new skin to protect us. Similarly, when soil is exposed in nature, the Earth naturally works to cover it. Grass, weeds, leaves, or fallen organic matter quickly blanket bare soil to protect it and promote healing.

This process reflects the Earth's inherent desire to sustain and regenerate itself. Which is very different than common farming methods, where we repeatedly expose and disrupt the soil. In modern agriculture, we grow crops, harvest them, and then reopen the Earth for the next planting, stripping the soil of nutrients and leaving it vulnerable. In permaculture, however, we allow the Earth to follow its natural rhythms… always covering, protecting, and enriching the soil.

Understanding this has deepened my respect for God's design for the natural cycles of the Earth. When we embrace these principles, we not only grow healthier plants but also align ourselves with the wisdom of nature, which has sustained life on this planet for generations.

Soil Intelligence

I like to think of soil kind of like an investment account for nutrients. When we grow plants, we "withdraw" nutrients from the soil. To keep the soil "profitable," we need to "deposit" more nutrients than we withdraw. If we don't, the soil becomes poor and lifeless, unable to support healthy growth. The good news is that everything we need to restore soil health is already around us. By composting, planting trees, and adopting smarter growing methods, we can rebuild our soil and make it healthier than ever before.

Investing in our soil isn't just about growing better crops; it's about creating a future where the soil is richer and more abundant for generations to come. Everything we need is here; we just have to be intentional and work together with nature to make it happen.

CHAPTER 5

SEED VS. SEEDLINGS – CHOOSING YOUR START

LEARNING HOW TO GROW FOOD has been far more rewarding than I ever imagined. Whether you're starting with seeds or seedlings, each option comes with its own unique challenges and rewards. For beginners, seedlings are often the easiest way to gain confidence because they provide a head start in the growing process. However, understanding how to grow from seeds is just as important. It's a skill that connects you directly to the natural cycle of reproduction.

This chapter will help you choose the best starting point for your garden. Along the way, we'll explore the principles of patience, persistence, and the beauty of nature's design. These principles remind us that everything we need is already here FREE.

What's the Difference Between a Seed and a Seedling?

Honestly, seeds blow my mind! They are beyond my comprehension. If I had to explain one, I'd say a seed is the very beginning. A small, magical capsule that contains what a plant needs to *become* that plant. Inside, it holds the blueprint for roots, stems, leaves, and eventually fruit. However, seeds are delicate and require specific conditions like the right soil, moisture, and temperature to sprout successfully.

A seedling, on the other hand, is the next step. It's a baby plant that has already germinated (sprouted) from a seed. Seedlings are typically grown in controlled environments, like greenhouses, and then sold to gardeners as a head start. When you buy a seedling, a potentially challenging part of getting the seed to sprout, has already been done for you.

If you're just starting out, seedlings are a fantastic way to see quick results and build your confidence. Here's how to choose healthy seedlings:

1. **Look for Strong Stems**: A healthy seedling will have a firm, upright stem. Avoid seedlings that look thin, weak, or floppy.

2. **Check the Leaves**: The leaves should be green and vibrant, without spots, holes, or discoloration.

3. **Inspect the Roots**: If possible, gently check the roots. Healthy roots should be white and well-developed, not brown, mushy, or tangled.

Starting with seedlings lets you skip the germination stage and focus on planting, watering, and watching your garden grow. It's an excellent way to begin while learning the basics of plant care. When I started my first garden, I chose to begin with seedlings. At the time, I didn't have the experience to grow from seeds, and starting with seedlings felt more realistic. I planted a variety of lettuce seedlings like butter lettuce, romaine, red romaine, arugula, and kale. What I loved about growing lettuce was how I could pick leaves from the bottom of each plant as they grew. The plants kept producing new leaves at the top, and my family enjoyed fresh lettuce for months. For a few months, we had more lettuce than we could eat!

But then something changed. The lettuce plants started growing tall stems, and eventually, flowers appeared at the top. The leaves became bitter, and through research, I learned the plants were "bolting." This is their natural way of reproducing. The flowers at the top of a bolted plant eventually turn into seeds!

During my first harvest of lettuce, I didn't realize this. I pulled the plants out of the ground shortly after they turned bitter, missing the chance to collect the seeds. But the next year, I let the lettuce plants bolt and collected the seeds. This year, I just planted those seeds for the first time, and I'm excited to see the results! Check out a video I posted while I was writing this book. Simply open your phone's camera, point it at the QR code, and tap the link that appears on your screen.

This experience taught me an important lesson: by allowing the natural cycle of reproduction to happen, plants can provide not only food but also seeds for the next generation. Starting with seedlings made things easier for me as a beginner, but over time, I've learned to appreciate the patience and rewards that come from seeing a plant through its entire life cycle and growing the next batch from seeds.

Growing from Seeds

Growing from seeds requires more attention to detail, but it's incredibly rewarding especially since one plant can produce hundreds of seeds! Here are a few simple tips to identify good seeds:

1. **Check the Seed Packet**: If you're buying seeds in a packet, look for details about the germination rate (usually listed as a percentage) and the best time to plant. A higher germination rate means a better chance of success.

2. **Inspect the Seed Appearance**: Good seeds are firm and don't have cracks. They won't be shriveling. Avoid seeds that look discolored or damaged.

3. **Try the Water Test**: This is my favorite test. Place seeds in a cup of water to check their viability. Seeds that sink are more likely to grow, while seeds that float may not sprout.

When planting seeds, it's important to provide the right conditions, including proper soil, moisture, and temperature. Some seeds need light to germinate, while others prefer to be buried. Whether you start with seeds or seedlings, gardening teaches us the value of patience and persistence. A seed doesn't sprout overnight, and a seedling doesn't become a mature plant in a day. It takes time, care, and consistency to nurture a crop

SEED VS. SEEDLINGS

all the way to harvest. The story of my lettuce garden is a perfect example. By letting the plants bolt, I discovered they could provide not only food but also seeds for the next crop. Now, I have more seeds than I can plant... and I didn't spend a dime on them.

This natural cycle of reproduction is a big part of why I titled this book *The More We Grow, the More We'll See Everything We Need Is Free*. When we let nature do what it's designed to do, we realize abundance is all around us. A single plant can produce hundreds of seeds, and those seeds can grow into hundreds of plants... creating ABUNDANCE! Both seeds and seedlings have their place in the garden, and both can lead to a lifetime of growth and learning. As you grow your garden, remember this... every seed holds the potential for abundance, and every plant is part of a natural cycle that can sustain you and your family for years to come.

CHAPTER 6

PROPAGATION, AIR LAYERING, AND GRAFTING – MULTIPLYING PLANTS WITHOUT SEEDS

WHEN I MADE UP MY MIND to become a student of agriculture and set a goal to master this space, I decided to document my journey. I started saying the phrase, *"I'm learning as I GROW,"* and later, *"The more I grow, the more I see everything I need is free."*

Along this journey, I've been amazed by the generosity of others who know much more about agriculture than I do. They share tips, encouragement, and support personally and in the comments of my videos. One of these individuals is Jayme, a fellow gardener I connected with through Instagram. He invited me to his property in Hollywood, South Florida, to exchange plants. When I visited Jayme's property, I was blown away. His backyard is an oasis of edible plants. Every tree served a purpose, whether it bears fruit or its leaves can be used for tea. It is a living example of the vision I've had for my own property. A beautiful space full of plants with purpose!

As we were leaving, Jayme pointed to a plant called the Mexican sunflower. He told me it was one of the easiest plants to propagate.

"Propagate?" I asked.

He explained that propagation is a method of growing new plants from parts of existing ones. All you have to do is cut a branch, stick it in the ground, and it will grow into a whole new plant. At first, I thought he was joking. But he cut a few branches, handed them to me, and encouraged me to try it at home.

In the last video about the lettuce seeds, I also showed how to propagate kale. If you haven't watched that video yet, feel free to check it out to see the process. What's really interesting is that the same kale plant I propagated in that video came from clippings Jayme gave me three years ago. This is the third generation of kale grown from those same clippings!

When I got back, I planted the branches outside my front door. To my surprise, it worked! Those branches grew into a sprawling Mexican sunflower plant, which is still thriving today. This experience sparked my curiosity and led me to explore other methods of propagation.

Air Layering – Growing a Tree Without Seeds

One of the most fascinating methods I discovered is **air layering**, a technique especially useful for fruit trees. With air layering, you can grow an entirely new tree from a branch, and it will produce the exact same type of fruit as the original tree!

Here's how it works:

1. Choose a healthy branch on a tree you want to propagate.

2. Remove a 2–3-inch section of bark around the branch, exposing the inner wood.

3. Wrap the exposed section with moist soil or moss and secure it with plastic wrap, tape, or a small container.

4. Over time, roots will begin to grow from the exposed area.

Once the roots are well-developed, you cut the branch just below the new roots and plant it in the ground. It's an incredible way to multiply fruit trees without needing seeds.

Grafting – Combining Plants for Desired Results

Another method of multiplying plants is **grafting**, where you attach a young shoot or twig of a plant (called a scion) from one plant onto the rootstock of another. The scion grows and produces the same type of fruit as the original tree.

The basics of grafting are:

1. Select a healthy branch from a tree with the desired fruit.

2. Make a clean cut on both the branch and the rootstock.

3. Join the two pieces together and secure them with tape or grafting wax.

Grafting allows growers to control the type and quality of fruit a tree produces, and it speeds up the process of getting fruit compared to growing from seeds. However, as incredible as grafting is, it led me to reflect deeply on nature and the original design of creation.

My Choice as of Now

Around the time I created *Jabez Mango Seeds* and set my goal of bringing over one million fruit trees to America, I also began reading the Bible more intentionally. As I studied Genesis, I noticed the mention of the Tree of Good and Evil, which was described as a "grafted tree." And everyone knows what happened to us after even Adam ate from that tree...

This stood out to me because I was just beginning to understand grafting as a technique. Many people encouraged me to focus on grafting for its speed and predictability. But as I reflected, I made a personal choice to focus on growing fruit trees primarily from seeds and using propagation and air layering sparingly rather than grafting. Why? Because growing from seeds feels more in alignment with the natural order of creation. A seed holds within it the of the greatest mysteries of life... Diversity!

I can't help but notice how plants, animals, and humans all share something remarkable: even within the same group, each individual has its own unique traits. Look at humans, for example, every single person is different. Even within families, brothers and sisters born to the same parents have their own distinct characteristics. It's the same in the plant world. This makes me wonder, was it not meant to be this way for a reason? Nature seems to have its own design, creating diversity with intention. So why do we, as humans, often insist on growing so many of the exact same kind of crops? Maybe it's time to step back and let Nature do what it's been perfectly programmed to do.

While grafting is a powerful tool, it's a form of human manipulation. I want to see what happens when I grow trees in reflection of the original design. Honestly, this decision comes with some uncertainty though. Growing from seeds takes much longer, and the exact type of fruit produced isn't always predictable. It may take 20 years before I fully understand the outcomes of my choice, but I'm willing to learn this lesson through patience and persistence.

Multiplying Plants for Free

One of the most beautiful lessons I've learned is that plants are capable of multiplying themselves in more than one way. Whether through propagation, air layering, or grafting, the principle remains the same... you can grow new plants without seeds.

For example:

- When you cut a branch from a tree and propagate it, the tree will grow back that branch.

- The new plant you grow from that branch can eventually be propagated itself, creating an endless cycle.

- This process costs no money and leads to a multiplication and abundance.

This is why I titled this book *The More We Grow, the More We'll See Everything We Need is Free*. Nature provides everything we need. We just have to learn how to work with it!

Lessons in Patience and Persistence

Propagation, air layering, and grafting all teach us the same principle: growth takes time and care. Whether you're waiting for roots to form, a graft to take, or a seed to sprout, you're practicing patience and persistence. These methods also remind us of the natural cycle of abundance. A single tree can produce countless branches, and each branch has the potential to become a new tree. This endless cycle mirrors the original design of creation. Everything we need is already here, waiting for us to use it wisely.

As you explore these techniques, I encourage you to find what works best for you. Whether you propagate, air layer, or graft, each method has its own beauty and purpose. And as you grow, you'll begin to see what I've come to realize… The more we **GROW**, the more we'll see **EVERYTHING** we need is **FREE!**

CHAPTER 7

FERTILIZER – NOURISHING THE ROOTS

THE ROOTS OF ANY PLANT ARE ITS LIFELINE. They draw nutrients from the soil, fueling growth and determining the plant's overall health and resilience. Fertilizer serves as a concentrated boost of nutrients for the soil, replenishing what's lacking or depleted to ensure strong, vibrant growth. My first experience with the power of fertilizer came during my second garden. My first attempt at gardening was on the side of my house, where I planted romaine lettuce.

I used a mix of organic soil from Lowe's and the existing ground soil. While the plants grew, they didn't produce much.

Determined to do better, I planted a second garden focused on growing lettuce and herbs. This time, I sought advice from a local nursery. When I asked about improving my results, the nursery worker recommended fish fertilizer. I had never heard of it, so I asked what it was.

With a chuckle, he replied, "Fish doodoo."

He then explained how to mix it with water, add extra topsoil to the garden, evenly distribute the fertilizer-water mixture across the soil, and let it rest for twenty-four hours before planting. Following his advice, I prepared the soil and planted lettuce and herb seedlings. The results were much better! The plants grew faster, healthier, and larger than in my first garden. The only major difference was the fertilizer. This experience taught me that sometimes, even the smallest addition, like nourishing the roots, can lead to extraordinary results. It also marked the beginning of my appreciation for the critical role fertilizer can plays in growing food.

FERTILIZER

That second garden taught me an important lesson: plants absorb nutrients from the soil as they grow, leaving less for the next round of plants. After my first batch of lettuce bolted and I planted new seedlings, I noticed the second crop wasn't as strong. This puzzled me at first, but the solution became clear: I needed to put nutrients back into the soil. Fertilizer became an answer.

Fertilizer comes in two main forms:

1. Natural/Organic Fertilizer: Derived from organic materials like compost, plants, or animal byproducts.

2. Synthetic/Chemical Fertilizer: Manufactured and often fast-acting but potentially harmful to the soil and environment over time.

I've always leaned toward natural fertilizers not just because they align with nature, but because they remind me of my childhood. Growing up, my Grandpa Lepak had a garden that produced the most incredible vegetables and fruits. I remember after fishing with him, cleaning the fish for dinner, he would have me wrap the fish heads, guts, and skins in newspaper, dig a hole in the garden, and bury them. He explained that the fish remains acted as fertilizer for the plants, enriching the soil with nutrients. He did the same with eggshells, vegetable scraps, fruit peels, and even stale bread. Instead of throwing old food away, it all went into the soil to feed the plants.

The garden soil was deep and rich, with a healthy texture that felt alive. Every summer that I visited, I was impressed at the garden's abundance, and it seemed like everything planted there flourished. Those simple lessons stuck with me. Natural fertilizer isn't just about adding nutrients, it's about working with nature making sure to feed the ground to create a cycle of growth and abundance. It's a partnership, where we give back to the soil so it can continue to give to us.

One of the most incredible natural fertilizers I've discovered so far is the **comfrey plant**. Its roots grow deep into the ground, pulling up minerals that most plants can't reach. These minerals collect in the leaves, which can be used as a potent fertilizer. I first came across the comfrey plant at a nursery called Ready To Grow Gardens in Miami. The owner Dylan told me it was an excellent fertilizer, and I got a cutting of the root. He explained that I could cut the root into chunks and each piece would grow into a new plant (another example of propagation: Root propagation). Check out a video I recorded that day about this. Simply open your phone's camera, point it at the QR code, and tap the link that appears on your screen.

I planted a few roots around my property, and now I have comfrey growing in multiple locations.

One day, a beautiful Bougainvillea plant in our backyard with purple flowers started dying. I had the idea to blend comfrey leaves with rainwater and pour it around the base of the bush to "feed" its roots. Within weeks, the plant came back to life! I also noticed our sugarcane plant was growing really slow so I used the same mixture on it, and it started growing faster! Check out a video I recorded about this experience. Simply open your phone's camera, point it at the QR code, and tap the link that appears on your screen.

Making Fertilizer From What's Around You

The beauty of natural fertilizer is that it's all around us. Here are a few simple, cost-effective examples you can try:

1. **Banana Peels**: Rich in potassium, banana peels can be chopped up and buried in the soil or blended with water to create a liquid fertilizer.

2. **Eggshells**: Crushed eggshells are an excellent source of calcium, which is essential for strong roots and healthy plant growth. Let the eggshells dry, then grind them in a coffee grinder. You can sprinkle the powder around your plants or add it to your compost.

3. **Vegetable and Fruit Scraps**: Instead of discarding these, compost them or bury them directly in your garden to add valuable organic matter to the soil.

4. **Grass Clippings & Weeds**: Grass clippings and weeds, like nettles for example, can be soaked in rainwater to create a nutrient-rich fertilizer tea.

5. **Fish Remains**: As my grandfather taught me, burying fish heads and guts directly in the soil is a time-tested method for enriching the ground with nutrients.

These natural materials are around us anyway. By using them instead of tossing them away, you can nourish your plants naturally while saving money on store-bought fertilizers. It's a simple and productive way to work with nature and give back to the soil. Fertilizer is a way of giving back to the soil, ensuring that the cycle of nutrients within it continues. There's no one-size-fits-all approach, so I encourage you to experiment to find what works good for your plants. The usage of fertilizer ties directly back to the philosophy of this book. Nature freely provides everything we need to nourish our plants and ourselves. From fish remains to fallen leaves and fruit scraps, what we once considered waste becomes a resource.

Fertilizer is just one example of how nature invites us to participate in its cycles, giving back to the soil so it can continue to give to us. By working with the natural process of nature, we not only grow healthier plants but also deepen our connection to the earth and its abundance!

CHAPTER 8

HARVESTING – WHEN AND HOW TO REAP WHAT YOU SOW

Harvesting is one of the most rewarding parts of gardening. Knowing *when* to plant and when to reap what you've sown can make the difference between a good harvest and one that falls short. The timing of your harvest isn't just about the growth of your plants… it's about understanding the seasons, your local climate, and the needs of each plant.

Timing Is Everything

When I began my lettuce and herb garden in the backyard, I planted the seedlings in January. Within a month, I was harvesting fresh, crisp lettuce leaves from those plants. It felt amazing to step outside, pick what we wanted for a meal, knowing it came from our garden. But as spring rolled in and the temperatures began to rise, the lettuce plants started to bolt. The stems grew tall, flowers formed, and the leaves became bitter. That's when I realized I could have started the garden in October instead. By planting earlier in the cooler months, I could have extended my harvesting period by at least a couple of months before the heat made the plants bolt.

This experience taught me the importance of planning your garden with the seasons in mind. Each plant has a preferred growing season and understanding this helps make the most of your harvest.

Every crop has its own rhythm. Some thrive in the cool months, while others need the warmth of summer. Remember planting expectations change depending on the growing zones. To illustrate this point here are a few examples of crops, their ideal planting times, and when you can expect to harvest them:

Cool-Season Crops:

1. Lettuce, Kale, and Spinach: These leafy greens prefer cooler weather.

    - Planting Time: Early fall (September/October) or late winter (January/February).

    - Harvest Time: About 4–6 weeks after planting. In cooler climates, you can enjoy multiple harvests before the plants bolt in the heat.

2. Carrots and Beets: Root vegetables thrive in cooler temperatures.

    - Planting Time: Late summer to early fall (August/September).

    - Harvest Time: 2–3 months after planting, before the ground freezes.

Warm-Season Crops:

1. Tomatoes and Peppers: These plants do well in heat.

    - Planting Time: Late spring (April/May) after the last frost.

    - Harvest Time: 60–90 days after planting, depending on the variety.

2. Zucchini and Cucumbers: These fast-growing vegetables also prefer warm weather.

    - Planting Time: Late spring (May/June).

    - Harvest Time: Within 50–70 days of planting.

Long-Season Crops:

1. Sweet Potatoes: These take their time but are worth the wait.

    - Planting Time: Late spring (May/June).

    - Harvest Time: 4–5 months later, in early fall.

2. Garlic: A unique crop that spans seasons.

    - Planting Time: Late fall (October/November).

    - Harvest Time: Mid-summer (July/August) the following year.

Timing your crops with the season ensures that you can harvest them at their peak. Planning your garden with this in mind not only maximizes your yield but also saves you time and frustration.

Turmeric Lesson

The first time I saw a turmeric plant, I was at a Ready To Grow Gardens nursery with the owner Dylan. As we walked through the rows of plants, he pointed to the turmeric plant. I had no idea what turmeric looked like above ground. I only knew it as a root that grew underground. The nursery owner explained that turmeric is simple to grow. You can take a piece of the turmeric root, plant it in soil, and it will grow into a whole new plant, producing a fresh bundle of turmeric roots. This fascinated me. One small root could yield an abundance of turmeric, and a portion of that harvest could be replanted to keep the cycle going. Check out a video I created that day at the nursery. Simply open your phone's camera, point it at the QR code, and tap the link that appears on your screen.

This discovery ties perfectly into the timing of root vegetables. Like turmeric, crops such as carrots, beets, and radishes thrive when planted in cooler months. For example:

1. **Carrots**:
   - **Plant**: Late summer to early fall (August–September)
   - **Harvest**: 2.5–3 months later, in late fall or early winter (October–December), before the ground freezes.

2. **Beets**:
   - **Plant**: Late summer to early fall (August–September)
   - **Harvest**: About 2–2.5 months later, in late fall (October–November).

3. **Radishes**:
   - **Plant**: Late summer to early fall (August–October)
   - **Harvest**: Quick-growing and ready in about 25–40 days (September–October).

Remember, asking ChatGPT and watching how-to videos on YouTube are extremely helpful for learning the specifics of planting and harvesting the crops you would like to grow. These resources have helped me understand the steps involved and how to time planting correctly. Again, I encourage you to take advantage of these information resources to support your growing journey!

Saving Some of Your Harvest for the Next Cycle

My experience with the turmeric plant is also a reminder of nature's abundance. I learned that day that all you have to do is bury a piece of the turmeric underground, and it'll grow a whole new plant producing another bundle of turmeric! By using a part of your harvest, whether it's a few pieces of turmeric, cloves of garlic, or seeds from bolted lettuce plants, you can replant and start the cycle again. This simple act turns one harvest into the beginning of the next, making it possible to provide food for life… for free!

THE MORE WE GROW THE MORE WE'LL SEE EVERYTHING WE NEED IS FREE

Harvesting is not just about reaping what we've sown; it's about understanding the cycles of growth and renewal. The more we grow, the more we can see that everything we need is already here... free!

CHAPTER 9

HEALING SICK PLANTS – DON'T BE SO QUICK TO GIVE UP

PLANTS ARE SOME OF THE MOST RESILIENT organisms on Earth. Even when they appear weak, droopy, broken or nearly dead, they often possess an incredible strength to bounce back; especially when given the right care. In this chapter, I want to inspire you to recognize that hidden strength in plants and show you how small, simple actions can bring them back to life.

When I was a child, between birth and age seven, I spent a lot of time in Wisconsin and Minnesota. I was fascinated by the trees. In the summer, they were full of life. Green, vibrant, and strong. But in the winter, they looked bare and lifeless, almost as if they were dead. I have a memory from when I was young of a tree on my grandparents' property that stood out to me. Unlike the other trees, this one stayed lifeless all year long. In the summer, when every other tree was vibrant, full of leaves, and teeming with life, this tree remained bare. My grandpa told me it was dead and that one day he would cut it down to use for firewood.

What stuck with me most was how different that tree looked in the winter. When the seasons changed, all the once-lively trees became still and bare, stripped of their leaves. Suddenly, the dead tree no longer stood out. It blended in perfectly with the others. The same trees that had been so full of life in the summer now looked just like the one that was truly lifeless.

This memory has stayed with me because it reminds me of an important truth: just because a plant looks dead doesn't mean it is. Many plants, like those trees in winter, may only be dormant, waiting for the right conditions to grow and thrive again. It's a lesson about seeing beyond appearances. Also, many plants that seem beyond saving can recover with the right care.

This chapter is about hope. The belief that even when a plant looks like it's on its last leg, the right conditions and attention can bring it back to life. With a little patience and effort, we can restore health to our plants and witness their resilience firsthand.

Our Bougainvillea and Mexican Sunflower Experience

There are two plants that taught me a powerful lesson: never give up on something that looks like it's beyond saving.

Take our purple bougainvillea I mentioned earlier, for example. When it first started dying, I didn't know what to do, so I reached out to the person who recommended the nursery where I bought it. She gave me simple advice: give it about a gallon of water daily, trim back the edges, and add fertilizer. I followed her instructions, and to my surprise, the bougainvillea came back to life!

A year later, the same plant began to struggle again. This time, I decided to take matters into my own hands. I pruned it back, used homemade comfrey fertilizer, and watered it daily. Sure enough, it bounced back stronger than ever.

The second lesson came from the Mexican Sunflower cutting I mentioned earlier in the book that I propagated. When I first planted the branch cutting into the ground, it started looking bad pretty quickly. So bad that it seemed like a lost cause. The leaves fell off, and soon after, the stem began browning almost all the way down to the ground.

But at the base, where the stem met the soil, there was still about an inch of green. That tiny bit of life gave me hope. I pruned the stem down to the healthy green section and began watering it daily. After some time, tiny green shoots began to emerge from the stem.

Seeing this reminded me that what looks dead often just needs a little care, patience, and the right conditions to thrive again. Now, that Mexican sunflower plant next to our front door is taller than me!

Why Plants Suffer and How to Help

When a plant starts to suffer, it's a sign that something the plant needs is out of balance. Here are five common reasons plants struggle and examples of how to bring them back:

1. Lack of Water: A lot it plants need moist soil to stay healthy. If a plant's leaves are drooping or curling, it may be dehydrated.

Solution: Adjust your watering schedule. Give the plant enough water to moisten the soil without oversaturating it.

2. Too Much Water: Overwatering can cause roots to rot, cutting off the plant's ability to absorb nutrients.

Solution: Let the soil dry out before watering again and check for good drainage in the pot or garden bed.

3. Nutrient Deficiency: Yellowing leaves or stunted growth often indicate a lack of nutrients.

Solution: Add a natural fertilizer like compost or options I covered in the previous chapter about fertilizer, to replenish the soil.

4. Pests: Small holes in leaves or sticky residue might mean pests like aphids or spider mites are eating your plant.

Solution: Use natural pest repellents, such as neem oil or homemade sprays using peppermint oil.

5. Sunlight Issues: A plant not getting enough sunlight can become **leggy**, which means it grows tall and thin with weak stems as it stretches toward the light. This makes the plant look unbalanced and can affect its health. On the other hand, too much direct sunlight may scorch its leaves, causing brown spots or wilting.

Solution: Move the plant to a spot where it gets the right amount of light for its needs.

HEALING SICK PLANTS

When plants show signs of trouble, I turn to ChatGPT to get advice tailored to my specific situation. Here's an example of how to ask the right questions and get helpful answers:

Question: "My tomato plant's leaves are turning brown and crispy. I live in Zone 10. What could be causing this, and how can I fix it? Please answer from the perspective of someone who has one over 10 awards for their expertise and achievements in agriculture. Please explain it in a way that is easy to understand using a step-by-step process."

Here is the answer ChatGPT gave: *Browning and crispy leaves on a tomato plant in Zone 10 can be caused by a variety of factors, but it's often related to environmental stress, watering issues, or disease. Let's go step by step to identify the cause and provide solutions:*

1. Check Your Watering Practices

Problem: *Both overwatering and underwatering can cause browning leaves.*

Solution: *Feel the soil about 2 inches deep. If it's dry, the plant needs water. If it's soggy, cut back on watering.*

- *Water deeply in the morning to ensure the roots absorb moisture and the leaves dry off before nightfall.*

- *In Zone 10, the heat can cause rapid water evaporation, so a consistent watering schedule is crucial.*

2. Inspect for Sun Scorch

Problem: *Intense sunlight common in Zone 10 can scorch tomato leaves, especially if water droplets linger on them and magnify the sunlight.*

Solution: *Use shade cloth during the hottest part of the day to protect your plants.*

- *Avoid watering in the middle of the day to prevent water droplets from acting like tiny magnifying glasses.*

3. Examine for Nutrient Deficiencies

Problem: Lack of essential nutrients, particularly potassium or calcium, can cause browning edges.

Solution: Feed your plant with a balanced fertilizer or compost tea rich in potassium and calcium.

Consider adding crushed eggshells or a calcium supplement to the soil to prevent deficiencies.

4. Look for Pests

Problem: Common pests like spider mites or aphids can cause browning by sucking nutrients from the leaves.

Solution:

Inspect the undersides of leaves for tiny insects or webs.

If you find pests, spray the plant with a mixture of water and a few drops of neem oil or insecticidal soap.

5. Inspect for Fungal Diseases

Problem: *Diseases like early blight or septoria leaf spot often cause browning leaves.*

Solution:

Remove any affected leaves immediately to prevent the spread.

Avoid overhead watering, as wet leaves promote fungal growth.

Apply an organic fungicide like copper spray or a homemade mixture of baking soda and water.

6. Mulch to Retain Moisture and Protect Roots

Problem: *Exposed soil can dry out quickly in Zone 10's warm climate, stressing the plant.*

Solution: *Add a 2–3-inch layer of organic mulch (like straw or wood chips) around the base of the plant. This helps retain moisture and regulate soil temperature.*

7. Prune Strategically

Problem: *Dense foliage can block airflow, leading to fungal issues or pest infestations.*

Solution:

Trim back lower leaves and overcrowded branches to improve airflow. This keeps the plant healthy and prevents moisture from accumulating.

8. Monitor and Adjust

Solution:

After addressing the issues above, monitor your plant daily. Look for signs of improvement or new problems and adjust your care routine as needed.

- *Keep a journal of what you've done and how the plant responds to build your gardening expertise.*

By following these steps, you should see your tomato plant bounce back. Tomatoes are resilient plants, and with the right care, they'll thrive even in challenging conditions like those in Zone 10. Don't be discouraged. This is all part of the learning process in gardening!

Asking ChatGPT questions like this will guide you toward solutions, but the real magic happens when you take action!

Check out this recent video I did about not giving up on a couple of fruit trees I'm growing. Simply open your phone's camera, point it at the QR code, and tap the link that appears on your screen.

A Seed for Thought...

One of the most fascinating discoveries in the plant world is that plants respond to care, not just physical care but emotional energy as well. Studies have shown that plants thrive in environments with kind words, positive emotions, and nurturing care. Conversely, plants in environments with negativity or violence most often show signs of stress. This connection between human energy and plant energy speaks to the deeper bond we share with nature. When we care for plants, we're not just helping them grow. I believe we're nurturing a part of ourselves.

As I write this, my goal of bringing over 1 million fruit trees to America through Jabez Mango Seeds is well underway. I've already given away 22 fruit trees. I we only have 999,978 more to go! But seriously, each time my wife and I plant a tree with someone, I notice something: every single person who has received a tree has had good energy. When we care for plants, we become caretakers of creation. It often starts with one plant, one garden, or one tree. From there, the care we show ripples outward, creating a chain reaction of positivity. Caring for plants not only nurtures them... it also changes us. By changing ourselves, we contribute to making the world a better place.

At its core, this is what the Naturalite Lifestyle is all about. I created the term "Naturalite" because there wasn't a category that specifically focused on breaking away from heavily processed foods and chemicals. For example, veganism focuses on avoiding animal products, but it doesn't address the consumption of heavily processed, chemical-filled foods. In fact, many vegan-friendly options are packed with additives and chemicals.

That's why I created the term Naturalite. A Naturalite is someone who actively chooses to move away from processed, chemical-laden foods by reading ingredient labels and selecting natural options. But the Naturalite Lifestyle is about more than just food. It's a mindset that embraces living in alignment with nature.

CHAPTER 10

THE MORE WE GROW, THE MORE WE'LL SEE EVERYTHING WE NEED IS FREE!

THOSE EMOTIONS IGNITED A FIRE within me. I made the decision to master agriculture and committed to learning everything I could about growing food naturally. It was a commitment I knew would be wise to keep for the rest of my life. The more I learn, the more I realize there's enough work to be done in agriculture to fill several lifetimes! It's a journey of growth not just in the soil, but in myself and one that I'm deeply passionate about continuing.

Early on, I thought it would be fascinating to document my journey online. I began sharing my experiences and lessons on Instagram, TikTok, Facebook, and YouTube. At first, it was simply a way to track my progress, but it didn't take long for me to realize that this journey wasn't just about me. As I kept posting, it became clear that my content could inspire others to grow food for themselves and their families. The more of us who grow natural food, the more we can create better options and stronger communities. This realization fueled my creativity and led me to ask an important question: *How can I make this content more valuable to others?*

From Johnny Appleseed to Jabez Mango Seeds

One day, while reflecting on how to bring more value to my content, I remembered the story of Johnny Appleseed. As a child, I was captivated by the idea of someone traveling around planting apple trees, leaving behind food for people and animals for generations to come. Around this same time, I made a new commitment to tithing 10% of my income. For me, tithing means giving back to the Creator of all things by serving people using the money I've been blessed with, as I believe people are made in the Creator's image. As my blessings increased, I began thinking of new ways to give back.

The memory of Johnny Appleseed aligned perfectly with the principle of giving, my content, and my tithing practice. That's when it hit me! *You can call me Jabez Mango Seeds!* I laughed out loud at how fitting it was. As I mentioned in a previous chapter, my mission became clear... to bring over 1 million fruit trees to America, starting right here in South Florida!

Jabez Mango Seeds: A Nationwide Naturalite Mission

Imagine a world where communities are connected through nature, where food grows abundantly, and where giving back to the Earth is a way of life. This is the vision behind Jabez Mango Seeds. I'm on a mission to inspire others to live in harmony with nature by planting and caring for fruit trees. Our goal is to give away over 1 million fruit trees, planting them on properties all over America!

This mission is an extension of the Naturalite Lifestyle mission which is built on the principle that *nature knows best.* The Naturalite mission is to ignite a global movement of consumption awareness—empowering individuals to choose natural food options over heavily processed, chemical-infused alternatives.

We believe true success comes from working with nature, not against it. Through ethical practices, education, and innovative products, we empower people to achieve their natural body weight, increase their energy, and reconnect with the wisdom of God's design. We are dedicated to inspiring communities to grow food naturally, restoring the connection between people and the land.

Our commitment goes beyond profits; it's about creating a legacy of wellness, sustainability, and integrity. By treating people with care, respecting the planet, and aligning every decision with what's right, we aim to build a valuable business that not only transforms lives but also sets a standard for doing business in harmony with nature. By aligning with nature, we not only nourish our bodies with better choices but also give back to the land that gives to us. Every fruit tree planted is more than a tree. It's a symbol of hope, a reminder of community, and a connection to the beautiful earth we are blessed to share.

Through this mission, my family and I have seen firsthand how nature has the power to bring people together, fostering patience, compassion, and unity.

Over the next five years, we plan to turn Jabez Mango Seeds into a nonprofit organization and expand our mission across the country. Our goal is to give away over 1 million fruit trees and plant them on properties nationwide. If you'd like to support this mission or have resources to contribute, please contact us at jabez@jabezinvests.com.

The More We GROW,
the More We'll See EVERYTHING We Need Is FREE!

The more I grow, the more I see that everything we need is already here. Nature provides for us freely and abundantly, but it's up to us to recognize and use these gifts wisely. By mastering agriculture, composting, and caring for plants, we can move from a mindset of scarcity to one of natural unity and abundance.

This is why I wrote this book. I believe that the more of us who grow, the more we'll realize how much is freely available to us. This shift in perspective can change the world; not just for ourselves but for future generations.

When we care for plants, we become caretakers of the Earth. That care brings out the best in us and inspires others to do the same. Together, we can bring back a world that is abundant, nurturing, and free!

Take the Next Step

This book is just the beginning of what's possible for you. We've created opportunities and resources to help you continue your journey and achieve even more. Whether it's joining a community of like-minded people, exploring other tools and products we've made available, or discovering what's next, we're here to support you every step of the way.

Simply scan the QR code on this page to explore everything we've put together for you. From contact information and exclusive content to products designed to make your journey easier, there's something here to help make a difference in your life.

We're excited to grow with you—because when you thrive, the world becomes a better place. Feel free to take a moment to explore what's waiting for you.

Pass It On

If this book has resonated with you, if it has opened your eyes to the abundance all around us and inspired you to take action, I encourage you to pass it on. Once you've internalized the lessons here and no longer need to reference this book, feel free to give it to someone else.

If there's someone you know who has the desire to learn how to grow plants (whether it's a neighbor, a friend, a family member) hand them this book and encourage them to start their own journey of growing natural food and herbs.

The more of us out here working with the elements, producing food for our families and communities, the better this world will be. Every seed planted, patch of soil nurtured, and every harvest shared is a step toward a better future.

Together, we can create a ripple effect. One person growing turns into two, then four, then countless others. This movement begins with you and it doesn't stop here.

THE MORE WE GROW THE MORE WE'LL SEE EVERYTHING WE NEED IS FREE

The more we grow, the more we'll see everything we need is free! Pass it on, and let's grow together!

Prayers Up! Blessings Up!
—Jabez

ACKNOWLEDGMENTS

In this section I would like to acknowledge the people that helped me in the creation of this book. Everyone on this list, I greatly appreciate for the role they played in guiding me through to the completion of my goals. If you are reading this and have a goal of writing a book, I'll leave contact information in case you would like to reach out to them. Trust me when I say they are all a pleasure to work with!

First, I want to thank God—the God of the Bible, יהוה, YHWH, also known as Yahweh. Coming into the knowledge of who you are and striving to walk in your ways has transformed my life in ways nothing else ever has. I am deeply grateful for the endless gifts you continue to provide, and I ask for your guidance to show me the path so I can serve you and the people you've called me to serve in a way that honors you and makes you proud.

Secondly, I would like to thank my wife, Nicole. To have a wife that makes a house a home is a MAJOR BLESSING. My wife Nicole does just that! She is an AMAZING woman, mom and my best friend. I love you, Bae!

I'd like to acknowledge Billionaire PA. I am truly grateful for the guidance I've received from him and for the friendship we share. He is truly a world changer. He taught me how to conceptualize, structure, and write a book—and also how to self-publish, own the rights to my work, and market it!

Contact him on Instagram: @billionairepa

Janine Hernandez is the final editor for this book. Not only does she do a great job, but she's a pleasure to work with. Her personality makes it extremely easy to communicate with her through the editing process!

Email contact: info@janinehernandez.com

I hired Tony Shavers to coach me on how to become a bestselling author, and he did just that! He walked me through the process step-by-step, smoothly and efficiently. If you have a goal of becoming a best-selling author, I highly recommend reaching out to him!

Email contact: info@tshavers.com

Jen Henderson at Wild Words Formatting designed my e-book and physical copy. She has a very structured process, and she does wonderful work!

Email contact: wildwordsformatting@gmail.com

Not only did Jamaal Wilcox design the cover of my book, but he also does all the digital designs for our business! This brother has a true gift for bringing ideas to life. If you're looking for someone to elevate your vision with graphic designs, I highly recommend him!

Email contact: malgrfx@gmail.com

THE MORE WE GROW THE MORE WE'LL SEE EVERYTHING WE NEED IS FREE

Lastly, I would like to thank YouTube, TikTok Instagram and Facebook for their platforms. Combined, they've helped me reach millions of people with my content!

My YouTube: @jabez_invest
My TikTok: @jabez_invests
My Instagram: @jabez_invests
My Facebook: Jabez Invests (If the page doesn't have over 400k followers it's not mine)

Made in United States
Orlando, FL
20 April 2025